Investing in the Imaging Supplies Aftermarket

To order additional copies, please contact us.
BookSurge, LLC
www.booksurge.com
1-866-308-6235
orders@booksurge.com

Investing in the Imaging Supplies Aftermarket

Demand-Builders, LLC

2006

Investing in the Imaging Supplies Aftermarket

Table of Contents

I
INTRODUCTION

Computers and printers are now established as basic communication tools. Typewriter ribbons have been replaced by inkjet cartridges. The cost of printers has declined sharply. The most expensive consumable is the ink cartridge and original equipment manufacturers (OEM) rely on the sale of cartridges as a significant profit center.

Just as in the automobile industry and many other sectors of the economy, a thriving "aftermarket" has evolved to offer a lower cost alternative to the OEM brand cartridges offered by the printer manufacturers. These "generic" brand cartridges are divided into two main groups, depending on the type of printer in which they are used. Some printers have the printing technology in the machine itself and are designed to use cartridges that are simply ink containers. Since there is no unique technology embedded *in the cartridge* itself, in these models, replacement cartridges can easily be designed and manufactured. These are brand new cartridges made to fit the printer but manufactured by someone other than the printer manufacturer. These are generally referred to as "compatible" cartridges.

Other printer manufacturers have elements of their printing technology embedded in the cartridge and their proprietary designs are protected by patents. Thus it is not possible to produce *compatible* cartridges without violating the patents. Under generally acknowledged principles of repair and maintenance, however, these cartridges, which are an integral part of the machine, can be refurbished and re-sold. These are generally termed "remanufactured" cartridges, i.e. they are used, empty cartridges that have been inspected, cleaned, some internal parts e.g. sponges, replaced, refilled and re-packaged. A sub-group in the industry simply drills a hole in the cartridge and refills it with ink. This latter group is generally perceived to be a lower quality product than the fully refurbished, remanufactured and filled cartridge.

Another aspect of this aftermarket is the *toner* cartridges which contain the toner ink used in laser printers. Laser printers are more common in the business sector, but the recent decline in the cost of laser printers is resulting in increased offerings of laser printers for the consumer market. Again, replacement toner cartridges for this segment can be either compatible or remanufactured. The key features of this segment are higher cost, longer life and follow-on support.

Without question, the printer supplies aftermarket is a substantial one. Using raw numbers, the profit potential is significant. But before investing in a distributorship it is essential that the prospective distributor fully understand the dynamics of the distribution model.

Our purpose in writing this book is to "level the playing field" between the companies that market distributorships and their potential buyers. The only way to effectively do this, in our opinion, is with information. The balance of this book is an attempt to pass on to these potential purchasers as much of our experience as possible so that they can make a fully informed decision rather than having to rely only on the information provided by the company selling the distributorships.

We urge you to read the book with an open mind; weigh the issues carefully; demand information and formal written responses to your concerns; and think carefully about the business model and the distributor's long range position in that model.

Most of all, we urge you to thoroughly investigate the market for independent, non-chain store sale of imaging supplies, the company that is marketing the distributorships (as well as predecessor companies) and the individuals who make up those companies. It is your first duty as an independent business owner, to be well-informed.

II.
DO YOUR DUE DILIGENCE !

a.   Sources

The most important thing you must do before you purchase a distributorship is a process generally referred to as "due diligence", in other words, conduct the type of inquiries that a person in the business would normally follow before agreeing to buy a company. If you have any doubt about the importance of this process, just look at the information that the selling company asks about you and your background. Are you able to obtain the same level of information about the company offering the distributorship?

You do not need an expert. What you need is someone who is not as excited or committed to the idea of purchasing a distributorship as you are. It could be a spouse, a friend, an advisor (lawyer or accountant) or a retailer with whom you are acquainted. You need this "impartial" party to balance the excitement and positivism that are characteristic of entrepreneurs. Once we are excited about a venture there is a danger that, like being in love, we will only see the good features of the venture and it will blind us to some of the potential problems or pitfalls.

The internet is a great resource. Search the net for any reference to the company you are planning to do business with as well as any of its affiliates, e.g. the locator company that the seller indicates will assist you in locating sites for cartridges. Ask how old the company is and if it previously did business under a different name or if it has recently been formed or if has recently purchased another company. Ask who you will actually be doing business with, in other words, many of these businesses conduct operations through a multi-layered system where Corporation A markets the franchises or distributorships, Corporation B is the "fulfillment house" that actually supplies the product and fills the orders either with its own brand of cartridge or that of Corporation A and Corporation C produces the remanufactured and/or compatible cartridges. Get all the names of individuals and corporations that you can and do a thorough internet search. Ask which state(s) the companies are incorporated in and where they are licensed to do business.

Contact the Better Business Bureau (BBB) in the city where the company is located to determine whether or not there have been serious complaints about the company and whether it has made good faith attempts to resolve those complaints. Do not rely exclusively, however, on the results of your inquiries to the BBB. If the company you are doing business with is new, the information is irrelevant. In addition, understand that the BBB is funded by the dues of its member companies and these bureaus can be slow to act on complaints against dues paying members. They also may have various standard responses, e.g. a company is "...under review", which permit them to delay giving any outright negative comment.

A new and controversial source is the internet based www. badbusinessbureau.com The Bad Business Bureau publishes something called the "Rip Off Report" where customers can vent their dissatisfactions with a company and its principals. This site contains complaints from consumers and others about local and national businesses. It appears that there is no rigorous pre-screening of complaints. Instead the company that is the object of the complaint is provided with an opportunity to respond to the complaint. While the complaints may be one-sided and biased, the site provides a counter-balance to the often glowing references provided by a company. In addition to consumer complaints, the site provides complaints by franchisees against franchisors as well as complaints by purchases of "business opportunities" such as ink distributorships. Remember to search under both present and past names of the company that you are contemplating doing business with as well as its subsidiaries. You can also search under the names of the principals of the company, if you are aware of them.

b.   References

Most companies will offer to put you in touch with existing distributors. Be sure to evaluate your contacts with these "references" very carefully. Just as you do when you apply for an employment position, one only offers as references those who have nothing but glowing opinions of your knowledge and skill. Some companies may offer concessions to key distributors as compensation for their time and effort in handling calls and questions from prospective distributors. Whatever the arrangement, the result is that these "references" are being compensated for their opinions. Don't expect the company to give you their list of distributors and allow you to choose anyone on the list. There are good reasons for the company not to do this. First, a distributor list is akin to a customer list. There are competitive and privacy issues. Second, the company is certainly within its rights to want to protect its distributors from an endless series of distracting calls from prospects. That doesn't mean you should not contact the proffered references. What it does mean is that you should carefully evaluate the information they provide. Make sure you "qualify" the reference so that you can put their responses in the proper context. For example:

- What is their name address phone number and business name?
- How long have they been a distributor?
- Have they always worked with this franchisor or supplier?
- What is the average number of cartridges sold per display per month?
- What type(s) of cartridges have the highest defect rate?
- What is the company's return policy on defective cartridges?
- What is the company's return policy on expired stock?
- Why did they agree to act as a reference?
- What types of locations work best for them?
- Where are their displays located e.g. urban, rural, stores, malls, kiosks?

Nevertheless, existing distributors remain the best way to evaluate the company you are planning to do business with. The problem is finding them. If you are in a city there is a good possibility that there are existing distributors already set up. An hour or two checking out likely locations can yield a triple benefit. First, you may find that there is a display already set up in which case the retailer will usually provide you with the information you need to contact his distributor. Second, the retailer can provide independent information about sales volume, types of cartridges that are selling, consumer reactions and questions and what their expectations are regarding service from distributors. . Finally, even if there is no display and no distributor information, the time spent is not wasted since you will have made an initial assessment of the viability of that location in the event you go forward with the purchase of a distributorship.

In addition, since you may be spending several thousand dollars or more why not consider an ad in the "business wanted" section of your local newspaper indicating your interest in purchasing an ink cartridge distributorship? For the $100 or so cost of the ad you might learn a lot about the industry, the business and perhaps even the company from which you are considering purchasing a distributorship. This small investment could save you thousands, could provide the opportunity to purchase an existing operation or, at the very least, could give you the assurance that you have thoroughly investigated the dynamics of the business you are about to enter.

c. Intermediaries—Who are you really talking to?

Many of the business opportunities today are the "brainchild" of one or a few entrepreneurs. In many cases they have opted to use independent contractors to grow their business as a means of limiting the demand for cash. Rather than incur the cost of employees and other features associated with large corporations, they have carefully outsourced many functions that only require them to pay for a service after they have made a sale. Alternatively, they contract with companies that specialize in finding potential buyers for their franchises or distributorships. While this is a perfectly legitimate procedure, and arguably a wise business approach, without careful oversight and supervision by the franchisor, there is a danger that the selling organization might overstate the opportunity or give inaccurate responses to questions about service or other aspects of the franchisor's business that are of no concern to them.

Therefore, in the course of your conversations with the marketer/offeror's sales force, remember to ask questions such as:

> Are you an employee of…..?
> How are you compensated?
> Where are you located?
> What is your connection with…..?
> Are you willing to put that in writing?

d.   Read the Documents

Once you come to a decision that you would like to pursue the opportunity, be sure to ask for, <u>AND READ</u> all of the documents that will form the agreement. DO NOT rely on any of the verbal assurances you may have received from anyone since the written documents you will be asked to sign will more than likely specify that the written agreement supercedes any verbal agreements. If an issue is important to you, ask that it be put in writing. And even if it is reduced to writing to your satisfaction, consider carefully what options will be available to you in the event the agreement is breached or the assurance ignored. Will you be able to get your money back? If so, what will you have to do and how long will you have to exercise that right?

Finally, recognize that you are stepping into the world of business. The company you are dealing with probably had an attorney prepare the agreement. Perhaps good business sense indicates that you should seek an attorney's advice and explanation of the documents as well.

e.   Outline of a Typical Agreement

Preliminaries

Recognize that it is THEIR document, not the typical contract that is negotiated between equals. The drafter of the document always has the advantage both as to what is in the document and, perhaps more importantly, what is <u>not</u> in the document. That doesn't mean you should not try to modify, eliminate or add items; just don't be too surprised if they refuse your requests.

- Parties—Be careful to note whether the parties to the agreement are the same parties that you dealt with.
- Exhibits and attachments—they are part of the agreement. Read them! Also this is how you might ask for something that you relied on in making your

decision be incorporated into the agreement, i.e. by making it an exhibit or attachment.

- Read the "Whereas" clauses. For example, one agreement says "WHEREAS, the company desires the Distributor to act as the direct interface with the *consumer..*" (emphasis added). That suggests that the company expects the distributor to market directly to the consumer; which may NOT be what most distributors contemplate since distributors deal mostly with retailers and it is the retailer who deals mostly with consumers. If you stumble on an issue such as that, demand clarification.
- Rights—most agreements will grant some kind of authority or right. It may be a territory or it might be only the right to sell the product. Read carefully.
  - o Sometimes these "rights" clauses also include obligations, such as that the company will be the sole source of supplies for the distributor.
  - o Sometimes it is more specific. For example it may be called a "requirements contract" obligating you to purchase only from the seller AND precluding you from buying or selling any OTHER products on your displays or racks, even if they do not compete with anything that the other party makes or sells.
- Territory—Note how it is defined. If it is with reference to roads or highways as boundaries, take the time to plot them out on a map. DO NOT assume that this is a protected territory unless it specifically so states. If it does not, recognize that the company is free to enter into agreements with others to have distribution rights to the same products in the same territory as yours.
- Term—Refers to the length of time the parties will have the rights and obligations as set forth in the agreement. Be aware of the automatic renewal

provisions, if any, and if you decide to proceed with the agreement, diary any preliminary notice requirements in your calendar so that the agreement does not automatically renew by oversight.

- Obligations—The agreement will invariably have a listing of Distributor Obligations. It will probably NOT have any such listing of the COMPANY'S obligations to the distributor.

- Ongoing operations—There will be some section of the agreement outlining the process for ordering shipping supplies. Take a calendar and plot out the timeline and procedures. Assume that it is six months or a year from now and your business is running along nicely. You place an order to fill the needs of your best retailer. Can you live with the timeline if the Company takes all the time allotted to fill your order. Recognize that under the terms of the agreement you might have to wait a month or more to receive your stock and that, under the terms of the contract, you cannot purchase stock from any other supplier nor can you cancel until the company's time limit for fulfillment, with any automatic extensions, has expired. What is the system for payment? Undoubtedly it will only be by credit card so that you, as a distributor, are denied the single most effective business tool for demanding performance, i.e. refusal to pay the invoice. Also, when and how will you be invoiced: When the order is placed?(whether or not product is immediately shipped) or will the company only bill you when they ship. Who pays shipping charges? (This will probably NOT be in the agreement as it is likely contingent on the minimum order size and the company wants the ability to increase the size of the order that qualifies for free shipping WITHOUT having to get your agreement to the change.]

- Relationship of Parties—In some cases you may be a franchisee, but in most cases the agreement will specifically state that you are not a franchisee and that the essence of the arrangement is that it is a "business opportunity". The franchising industry has been around for a long time and the franchise model is the predominant model for rapid expansions through the sale of rights to offer the franchised product or service. The by-product of this history is that there is a body of legislation as well as court decisions regulating the interstate sale of franchises. This also means that people who are contemplating purchase of a franchise must be given certain information and those who purchase a franchise have certain rights. Unfortunately, "business opportunity" offerings are not subject to the same degree of regulations and therefore, purchasers of business opportunities have fewer legal rights and recourses against the offeror of the business opportunity. As a result, nowhere is the maxim "Buyer Beware" more appropriate.

- Insurance—The Company will recite that it will maintain product liability insurance. It is not uncommon in the business world for parties to the agreement to request evidence of such insurance which request can be easily complied with by the company by having its insurance agent provide a CURRENT certificate verifying that the insurance is in force and the premiums are up to date.

- Warranties—This is a critical aspect of the inkjet cartridge aftermarket and it is to be expected that it will be clear and unambiguous as to the statement of quality of the product, the length of the warranty and the cure, i.e. money back or simply replacement. What starts the warranty clock running? Is it the sale to the end user? It may only be from the date it is received by the Distributor, meaning if sold more than one year after receipt by the distributor, the company is not obligated to replace a defective cartridge. A year seems like a long time but, what if you receive cartridges that only have six months left before expiration? Since the company will probably not accept expired stock, you only have a six month warranty on that particular cartridge. Also note the provisions, if any, dealing with the return of defective cartridges. Most agreements will provide that returns are subject to approval by the company. Look for re-stocking fees for return of overstock. Don't assume that you just box your returns up and ship them back. The agreement will probably specify that you have to get the Company's approval to return anything.

- "Freshness of stock"—If your business model is one of consignment then you bear the risk of expired stock. Look for something in the agreement that obligates the company to ship stock that has a minimum shelf live, e.g. one year. If it's not in the agreement ask to have it included. (You probably won't get it, but ask anyway.)

- Assignability—The agreement will probably state that the agreement can't be assigned to another party without written consent. That means that if you sell your business the buyer doesn't get to be the new distributor unless the company agrees.

- Breach—The only way to understand these provisions is to read through them twice. The first time, assume that you have violated the agreement. What are the rights and risks enumerated?. Then read through the provisions again but this time assume that you want to continue the business but the Company violated the agreement. What rights and risks do you have in that situation?

- Termination—Do the same exercise as you did for the Breach provisions.
- Indemnification—According to Webster indemnification means "..to compensate for damage or loss sustained". It may also include the obligation to hire counsel for the other party. These provisions attempt to describe the obligations or liabilities of each party to pay for claims and the cost of defending them.
- Jurisdiction/venue/applicable law—You can assume this section will select places most convenient to it.
- Binding—This is not the same as assignability. Usually it will provide that the agreement is binding on successors but not assignable. Again think this through from both perspectives. You give the business to your child or the Company is purchased by another company. Where do you stand?
- Other—Every such agreement will have some statement to the effect that "This written agreement is the entire agreement between the parties and NOTHING said or written by ANYONE, but not contained in this written agreement, shall be of any force or effect." This is probably the single biggest source of confusion and angst for people purchasing distributorships. The dynamics of the distributorship selling process are extraordinary. Phase One focuses on the "opportunity" to make money. Phase Two stresses how easy it is to operate the business and how many others are successful at it. As with any sales effort, both phases are also designed to foster a trusting relationship between you and the person or company marketing the "opportunity". Phase Three involves shifting the focus from trying to get your money to determining whether you will "qualify" for this opportunity. This shifts your attention from evaluating them to proving

your worthiness to qualify for the "opportunity". Phase Four is notification that the company will "accept" your money. Throughout all of these phases you have asked questions and received, presumably satisfactory answers to the point that you are now poised to sign the Agreement. You should treat this "...entire agreement" provision is a warning to you to IGNORE EVERYTHING YOU HAVE HEARD OR READ REGARDING THIS OPPORTUNITY. You are about to pay over your hard-earned money based solely what is contained in this agreement. If you are relying on anything else you have learned in the previous Phases, you do so at your own risk since, by signing the agreement you are agreeing that you were not induced to by the "opportunity" by anything and everything NOT SPECIFICALLY CONTAINED IN THE WRITTEN AGREEMENT.

   f.   Recognize That You Have No Leverage

After one finishes reading all of the various documents, it is easy to conclude that this is a binding legal arrangement in which both parties have rights. While that may, in fact, be true on paper, the reality of business life, and especially in these types of relationships, is that you the distributor will have little or no power to exercise over your supplier.

True you have an agreement, but you have already agreed that if you have a dispute it has to be litigated in the state where the supplier is located. You'll have to hire a lawyer in that state to assert your rights.

Agreements are only as stable as the company behind them. What if the company decides to sell the business? Where do you stand? In addition, most of the critical issues in your relationship with your supplier are PROBABLY NOT included in the agreement, such as, price changes, minimum orders, stock returns, re-stocking charges for slow moving stock, website maintenance costs, packaging changes that might require changes in your promotional costs, etc.

As with all companies, your supplier will do what is in its best interest and the expectation is that you will do the same. The only problem is that you are locked in to dealing with the supplier while your customers, i.e. your retailer and business accounts, are not locked in to dealing with you.

The purpose of these comments is not to dissuade you from going forward. Rather, it is to caution you to go forward with your eyes open to the realities of business and to suggest to you that, more than anything else, all you can rely on is the PAST BEHAVIOR AND REPUTATION OF THE COMPANY THAT YOU ARE DEALING WITH. The essence of the proposal you are considering is the PROSPECT (and specifically not the promise of) future income. Remember that a company's track record of selling distributorships is not the same as their distributors' track records of selling cartridges. It is the latter that is most important to you and it is doubtful that you will ever get any accurate information on the latter from the company selling distribution rights. A referral to one of their hand-picked best distributors is NOT data, and is NOT accurate information. It is merely a testimonial from one person who has done well. There may be hundreds or thousands of others who are doing very poorly but don't look to the seller to tell you about them or give you their telephone number!

III.
UNDERSTAND THEIR BUSINESS MODEL

Listen carefully to the supporting arguments advanced by the Company's sales staff. Are their justifications limited to the anticipated return from the primary business, i.e. the retail display racks, or do they also focus on the opportunities for "business to business" sales? Most consumers do not have laser printers where the toner cartridges sell for $50 to $100 or more each. However, these predominate in business and the repeat business is substantial and far more reliable than consumer retail sales. If the display distributorship is seen as just an entrée into this more lucrative business, then you should recognize that far greater amounts of your time and talent will be needed to be devoted to selling. And that will require a very different type of selling from seeking to place consignment stock display racks in retail sites. You will be working days, the decision makers will be harder to reach, the competition will be far more sophisticated and may involve features that require more of your time, e.g. pick up and recycling of used cartridges, on site inventory control responsibility, or the sale may be tied to elements you do not have the ability to provide, e.g. servicing of printers.

Finally, from the perspective of the Company, they are in the business of selling distributorships first and cartridge supplies second. The "big money" is in the sale of distributorship "rights". In order to determine the cost of these rights, ask for the cost (not selling price) of the stock on each display that is included as part of the new distributorship. Subtract the cost of the contents of all of the displays from the cost of the distributorship. The difference is the cost of the distributorship rights (assuming, of course, that you will ultimately sell every item of stock on all displays before the stock expires!). The value of these rights are grounded on two assumptions: exclusivity and inherent value. In many cases, however, there is no assurance of exclusivity and until you, or the company you are dealing with, has a valuable "brand" there is little or no 'inherent value" in those rights.

In most instances you will only receive a non-exclusive territory. That means that while you will have the "right" to sell the company's products in the defined territory, the company is free to sell those same rights, and products, in that same territory to others.

Think of a famous name "brand" such as a restaurant chain, auto repair franchise, or line of beauty products. These names or "brands" have inherent value based on long-standing presence in the market, years of advertising, repeated sales to customers and a host of other attributes. At this point the generic <u>brands</u> of aftermarket printer supplies have little or no inherent value. That will be your burden to carry and, unlike Staple® and other established brands that have introduced their own line of cartridges, you have no established brand that you can "extend" to this new line of products to enhance their credibility in the marketplace. In many cases your "brand" will, at least initially, be tied to that of your retailer.

In this regard, one of the questions you should ask the company selling the distributorships is "How much have you spent in the past on advertising to develop brand awareness of your cartridge line and how much have you budgeted to spend on national promotion in the coming year?" Don't settle for generalities in the answers to these questions. Ask for dates and media of past spending.

The company's business model may only be quick upfront cash with little attention or interest in the ongoing support and supply functions.

IV.
UNDERSTAND <u>YOUR</u>
BUSINESS MODEL

The distributor is the classic middle man. The Company has little risk because they are selling intangible rights for which they are fully paid up front. The assume no risk from ongoing operations because they only do business with distributors by credit card, in other words, they are ALWAYS PAID BEFORE THEY HAVE TO PERFORM, i.e. ship product or host your website for the following month. Even in the case of returns for slow sales or defective products, all you will likely receive is a <u>credit</u>; and a credit is only of value to the extent that you continue to purchase product from that supplier.

You on the other hand bear all of the risk. You sell on consignment so you place product without payment. Your products are sold without your contemporaneous knowledge by your retailers who may or not pay you at a later date. You bear the risk of expired stock. You have the burden of seeking to recover for defective stock from the Company. You may be forced by some of your retailers to bear a portion of the losses due to theft. Finally, in an era when the internet is constantly undermining the role of the middleman, you bear the ultimate risk that the market will seek to buy direct. [In this regard, it may be well to read the proposed Agreement to see if there is any undertaking by the Company that it will only sell through distributors. If not, you would do well to assume that the Company may some day find it more attractive to deal directly with a major customer and thereby recover a greater profit from the sale.]

Another aspect of your business model is the fact that it is single product (i.e. cartridges) retailing. As obvious as this sounds, there are significant implications for you as you seek to grow your business. Multi-product stores can amortize the cost of advertising over a wide range of products of interest to consumers, e.g. clothes, house wares, electronics, auto supplies, toys, etc. These major stores can recover the cost of their advertising in the first few sales depending on the product(s) purchased.

You, however, have a single product line each of which will produce $2 or $3 gross profit for each sale and which will be useable for one or more months. That means that you may need to make tens or hundreds of sales simply to recover the cost of the advertising.

Your business model is also based on distribution rather than direct sales. That means that you do no control most of the aspects of the retail sale, e.g. the selling staff, the hours of operation, all of which limits your ability to derive the maximum return on your advertising dollar.

Finally, remember that your business model is one of consignment stock on displays in leased space in retail stores. That means future expansion will tie up several hundred dollars for each new display that you place, depending on how extensively you stock each display. That makes these distributorships particularly capital intensive for the distributor, quite apart from the initial investment.

V.
LOCATION ISSUES

---

a.   Locator Services

Most "business opportunity" marketers (i.e. "Companies") will enhance their offer by including "free" location services. For example, if you are purchasing a package of three displays, they will agree to obtain agreements to place your displays in three retail locations in your area. This is very helpful, especially if you were attracted to the offer of being only a distributor ("...No selling required!"). It is important, however, that you understand the true value of that "free" service.

First, it may not provided by the company's staff. Rather the company contracts with another company that specializes in finding retail locations for distributors. Second, since the wholesaler is probably buying this service in bulk, it is likely that the service provider is getting a reduced fee. That may or may not affect how promptly they travel to your location. Third, it is even possible that the location service is a subsidiary of the company or has some other relationship with the location finding company. While this may be an asset, it can also be a liability since the location service has a locked in customer. In other words, you are not the location company's customer; the marketing company is their customer, not you. Further if there is a pre-existing relationship between the two it often means that the basis for the selection of the particular locator company is not focused solely on satisfactory performance as measured by the distributor since the locator is guaranteed the business by the marketing company, no matter how poorly they perform.

Do not assume that the locator's timetable will correspond with yours. The locators travel throughout the country and may be under contract with several different types of business opportunity or franchise marketing companies. It is more than likely that they will not drop everything to come to your city if you are the only new distributor. Rather they may tend to wait until there are several contracts to fill in that city in order to get the best return on their expenses. Consider also, that while it may be a locator company, the probability is that it has only one or two employees and operates exclusively through a network of independent contractors who are free to decline work that does not appear sufficiently lucrative. Some of these folks are expert at what they do. Others are in it for the freedom from supervision. They may or may not have a good feel for what is the best type of location for cartridge displays and, as a new distributor, you may not yet be sufficiently experienced at that point to effectively question their decision.

What does all this mean? One possibility is that everything goes like clockwork and 30 to 60 days after you have purchased your distributorship, you have stock, sites, and your first few cartridge sales. Given human nature however, as well as the disjointed relationships between the various parties involved, and the absence of traditional management controls over all aspects of the operation, it is more than likely that you will be sitting on your stock and displays, (with the expiration date time clock on each cartridge running) for several months before you have set up your first display.

b.   Doing it by Yourself

If you choose to expand your network beyond the initial complement of displays, you can choose between using a locator service which will charge a fee of perhaps $200 per location or you can take on the task yourself. There is much to recommend doing it yourself. First there is the cost. Assuming your average gross profit on each cartridge sold is $2 to $3, that means you will have to sell 70 to 100 cartridges from that site to recover just the locator's fee. Since you are in an expansion mode at that point, you should have a good idea about just how long it will take for you to recover that cost. Second, as discussed above, you are probably better motivated to evaluate and more knowledgeable about the area than some locator who drives in from another state. Finally there is the trade off between the locator's "experience" in finding sites specifically designed for selling cartridges versus the opportunity to look at new venues that you might bring to the task. Remember, the locator is going to favor sites that are "easy sells" to take the display. They may or may not be great retail locations for cartridges. Many retailers are attracted to the cartridge display because it expands their "brand" as a full service provider. Assuming space in the retailer's site is not at a premium, they are often not as concerned with the volume of sales as you are since they are deriving additional benefits from the presence of the display alone. While it is true that you may have the right to approve a site, rejecting a site means more delay and lost sales.

c.   Competition

1.)   "Big Box" Stores
The ink cartridge market is extremely competitive. The major office supply stores such as Staples, Office Max, Office Depot and others carry extensive cartridge inventories. Some even offer guarantees in the event they do not have a cartridge in stock. Some of these stores

have their own "house" brands of cartridges at prices close to what you might offer. In addition to extensive inventory, these companies have invested millions of dollars in building brand equity in their corporate names. Assuming price parity, most consumers will opt for a "name" vendor over an unknown.

Before you invest your hard earned money, spend an hour or two on the internet. Visit the websites of these major stores and use the location finder pages to plot the stores in the areas you are considering. Look for areas where there are no mega-stores within10 to 15 miles. These areas may be your best opportunity despite the fact that they lack the population density sufficient to be attractive to the major office supply chains. Of course, the lower population density in these areas also means lower sales volumes.

2.)   Price Issues

Next comparison shop the prices of the non-OEM brands sold at these stores. Unless you will have a significant price advantage you will be at parity with these stores on price. They might nevertheless have an advantage over you because of the value of their "name"; their hours of operation (how many people are able to purchase only between 9 am and 5 pm on weekdays?), which frequently include nights and weekends; their incentives (e.g. free paper with return of used cartridges) and the wider range of cartridges.

d.   Types of Locations

Low priced, generic inkjet cartridges are intuitively a great idea. Almost everybody you talk to will comment on the rapidly declining prices of printers and the ever-increasing cost of cartridges. The marketing principle at work here is the razor blade model: give away the shaver to get folks hooked and then make your money on the recurring blade sales.

The company trying to sell you a distributorship or franchise will include in their marketing materials a list of types of stores that are good prospects or where other distributors have reportedly been successful. It is critical that you make an independent assessment of these types of potential locations before you agree to buy. This can be accomplished very easily; simply visit each type of store in your area and talk with the manager. The following are some aspects that you may not have considered.

Chain stores, e.g. drug or convenience, gas stations
- Local managers may not have authority to agree to accept display;
- Buying power may require more demanding terms, e.g. greater commission; first turnover of stock free
- Product credibility—consumers may question the quality of product sold next to potato chip display
- Unwilling to assume risk of theft losses
- Requirements for product bar-coding

Packaging and shipping stores
- Lack of local authority to contract;
- Limited operating hours
- Reluctance to learn about product

Computer sales and service stores
- Hours of operation-(most rely on service fees not retail, for the bulk of their business)
- Some may even close in order to provide service at an important customer's office
- Defective cartridges reflect negatively on their brand image for quality.

School bookstores
- Many are operated by national chains such as Barnes & Noble®, and local managers do not have authority to contract

- Uneven demand due to class schedules and financial aid timing
- Use of school printers may be included in student fees thus reducing need to purchase cartridges

Print shops
- Core business is printing not retail;
- Concerned with reputation for quality (see "Returns" section)
- Limited hours

Local supermarkets
- Unwilling to assume risk of losses
- Issues regarding bar-coding and scanners
- Lack of interest in developing product knowledge

VI.
ADVERTISING AND
MARKETING

Putting a display in a store does not guarantee sales. Think about your own experience. Have you ever taken something in to a store for repair? Were you really "shopping" or were you on a single-purpose mission to get the item repaired? Can you even recall what other items were on the shelves? It is true that, over time, the retailer might build up a following of folks who rely on his or her store for inkjet cartridges. However, if the items are in the store on consignment the retailer may lack motivation to "push" the item since he has no financial investment in the product. More importantly, can you afford to wait that long to get the word out about your product? Remember, you already have your money tied up and that cartridge expiration date clock continues to click.

So eventually you will come to consider advertising. Here are some issues to think about.

1. Advertising is expensive and is priced on the basis of an assumption that everyone targeted by that particular medium, e.g. subscribers or passing cars, sees the advertisement. But the facts are a) not everyone sees the ad, b) only a fraction of those who see the ad remember

it, c) only a fraction of that sub-group agree with it and d) only a fraction of that sub-group are motivated to act on it. Therefore, before you buy your distributorship call your local free "weekly shopper" newspaper and get the rates for an ad. Then convert that cost to the number of cartridges you will have to sell to recover that cost (using the average profit per cartridge). Do the same for other advertising media you might be considering.

2.   Cooperative advertising—If you can get the retailer to participate in the cost of the ad, it can cut your costs and possibly make advertising a more viable option. You will, of course, have to agree on ad copy, placement, frequency.

3.   If your stores are only open weekdays from 9 to 5, does it make sense to spend money advertising if most people can only shop for cartridges after 5 or on weekends (i.e. when the mega stores are open)?

4.   Couponing—offering money off with coupons can be an effective way to stimulate business. You will need to decide whether you alone will absorb the cost of the reduced price or whether the retailer will participate. There are accounting issues to think about (How do you know which type of cartridge went with the coupon?) In addition, you need to design a coupon and determine how you will distribute them. Putting coupons on the retailer's counter may increase (discounted) sales but it doesn't bring anyone new to the retailer. Supermarket register tapes offer another alternative but you will have to consider timing (they might work only in 3 month segments and require advance notice). Again, your retailer's limited hours of operation may frustrate sales opportunities..

5.   Are you selling or distributing?—If you are getting into this business solely because you believe that all you will have to do is refill displays you may be in for a shock. True, there may be some routes, e.g. bread routes or

soda machines, where all you do is buy from a wholesaler, collect retail sales and restock, but eventually it all comes down to sales.

6. Relationship Selling—At a minimum you'll need to establish an effective relationship with your retailer. That relationship will be built on your readiness to obtain unusual cartridges not included in the standard mix or some other special request. Sooner or later the retailer will present you with cartridges that the consumer returned (opened!) because they mistook the type of cartridge they used. Will you replace it at no charge even though it was not defective? Finally, if you are ever going to have your display do anything other than collect dust, you are going to have to learn the dynamics of the retailer's business, as well as your own. Can you help him sell his core business; can you provide referrals; what type of signage can you offer that will enhance the store's image?

7. Site locating—As mentioned elsewhere, it is likely that you will find yourself looking for new sites, either because an existing site is not productive or because you wish to expand the number of outlets. Consignment sales are perhaps easier because the retailer does not have to make a cash outlay for stock. Nevertheless, most of the retailers you will deal with are small one-owner shops that have worked hard and long to build up a reputation in the marketplace. As you market to them, they will have questions about the quality of your product, how frequently you will visit the store to maintain and restock the display, pricing policy, payment policy and a host of other questions. Remember, you will not be there when the consumer considers a purchase. The retailer will be responsible for answering questions and vouching for the product. If you can not "sell" the retailer on all facets of your business, and especially your supplier's product, it is not likely that they will agree to participate.

VII.
TIME ISSUES

Do not assume that you can run your distributorship in your spare time outside of normal working hours. Unless you are lucky enough to find 24 hour outlets where the decision-maker is present after normal business hours, you will need to be able to meet with retailers during normal working hours. Be prepared for a number of cancellations. At that point you are just one of dozens of salespersons who are importuning the retailer every day. For example, when faced with a choice between an opportunity to fix a computer at their regular $30 to $70 per hour rate or to honor an appointment with a salesperson, the successful computer sales and service business owner will opt for the former without a second thought.

Presumably you have already recognized that as a small business owner you are required to keep reasonable business records and this will demand some of your time.

Depending on your particular style, you can probably assume that inventory checks, re-stocking, display maintenance (e.g. dusting, stock rotation) will run approximately one hour per site. Inventory duties such as re-pricing and removing price stickers before returning slow moving stock for credit will also usurp some of your time.

It is important to understand that, while companies may offer varying degrees of support, in most cases it will fall to you to design forms to be used in your business, signage, reference tools, etc.

Since the cost of traditional advertising may be prohibitive, it will be necessary to devote time to personal marketing and advertising. This may mean producing and distributing inexpensive flyers, developing a fund raising program based on return of empty cartridges, or similar time-consuming activity. While there is certainly room for innovative approaches to low cost marketing, the fact is that, if you do not buy advertising, most other forms of marketing and promotion will be labor-intensive on your part.

VIII.
INVENTORY CONTROL

Ask any accountant and he or she will tell you that inventory is one of the most difficult aspects of any business. In addition to so-called "shrinkage", there are multiple opportunities for errors in counts from orders to packing slips to invoices to stock control. In addition, as a company that sells on consignment, you will need to develop and understand a system to keep track of inventory and sales in order to make sure you have sufficient stock on hand to meet the needs of your retailers. This is slightly complicated by the fact that you will have, in effect, two separate inventories. Since the consignment items remain yours until sold, one part of your inventory is the stock on each display. The second inventory is the stock you maintain at your site and that you use to restock your retailers' displays.

Inkjet cartridges also require some attention with respect to storage, i.e. they are temperature sensitive. They have to be protected against freezing or extreme heat. That means that you cannot assume that you can leave them exposed in the sun in a hot vehicle in summer or that you can store your stock in an unheated garage in winter. It is also recommended that they be stored upright. Therefore you can not assume that you will be able to leave the cartridges in their shipping containers until needed.

As you might expect, inventory can be subject to sudden swings brought on by significant changes in they types of printers used. For instance, a major marketing effort by one of the printer manufacturers of a new type of printer might, in a few months, trigger a sudden spike in demand for a particular cartridge that you have not carried in the past and/or result in a sharp decline in demand for cartridges for a formerly popular printer.

Note also that your supplier will make changes in the cost of cartridges and recommended selling prices. This may require changes in your inventory accounting. Further, since most cartridges have expiration dates, your inventory control system for your facility as well as for each display, will have to be organized to sell the oldest, unexpired stock first.

Finally, you will also need to keep tract of slow moving stock so that it can be returned to the supplier for credit (less restocking fees) well in advance of the expiration dates.

IX.
CASH FLOW

As you mull over the possibility for expansion take time to compute the cost (at cost to you) of a fully stocked display. That is the amount of your cash that will be tied up at each site. Also note that there will be a considerable delay between the time you purchase stock and the time you are able to collect income. To be sure, your supplier will NOT extend you any credit. Everything you order will be billed to a credit card which you will be required to supply.

On the receivables side, however, the picture is much different. First, the retailer receives the full amount of the sale and has the use of the money until such time as you collect. If you intend to collect once a month, then this 30 day "float" is the retailer's. What if you are out of stock and cannot refill an item on the display? You will need a system to prevent double billing. (One approach is to only bill for items sold that you are able to re-stock. While this has the benefit of avoiding confusion, it also means that you do not get paid for a sold item until you are able to re-stock it, thereby reducing your cash flow.) Further, don't be surprised if a few of your retailers inform you that because of cash flow problems of their own, they are unable to pay you for the stock sold. [That's right! It's your money and they used it to pay other bills!] They will ask you to come back in a week or offer a post dated check and ask you to hold it for a week. Alternatively, they will write you a check only to have it returned to your bank for non-sufficient funds and your account charged $25 or so for the bounced check.

Stock maintenance can also impact cash flow. For instance, assume you want to collect and return a group of slow moving cartridges that will expire in three months. If you remove these cartridges from the display and return them for replacements, you will not have any to sell until the replacements arrive. This collect-inventory-pack/ship-receive replacements cycle might take 60 days or more. If you decide to order replacements first, you will have a cash outlay for the new stock. While the company will eventually issue a credit for the qualifying returned stock (less a restocking charge) this will only be a credit against future purchases.

X.
EXIT STRATEGY

Those promoting the sale of distributorships may seek to dismiss questions of exit strategy as inappropriate, or defeatist, or inconsistent with the entrepreneurial spirit they are looking for. Nevertheless, there may come a time when your situation changes due to relocation, health or other reason and it becomes necessary for you to sell the business. The time to consider this is BEFORE you buy the distributorship. Think of it as similar to a pre-nuptial agreement. Start by asking the seller of the distributorship rights what its policy is regarding resale of distributorships. Some may have a program to buy them back, probably at a discount. Others may offer to assist in finding a buyer. Still others may take the position that they will not assist in any way. (Generally, this is because they receive no income from the resale of a distributorship. It's the sale of new distributorships that drives their business model!) Whatever the policy, try to get it in writing and included in your purchase agreement. But remember, any agreement is only as good as the company that stands behind it and you will have the burden of the expense of enforcing those rights if the company chooses not to abide by the agreement.

One of the reasons why the sale of NEW distributorships is so successful is that what is being sold is POTENTIAL. And as a new prospect you are likely to see only the great potential for sales and success because you are committed to making it so. The seller has a clear advantage in this situation. When you go to sell your business, however, you are no longer selling potential. You have an <u>actual history</u> of sales and profitability, i.e. a track record and that gives an advantage to the buyer. Unlike you, who had to rely in great part on the representations of the business opportunity "marketer" or franchisor and its "references" as to the possibilities for income, your potential buyer can simply ask you for your sales and income statement. Further, if you have been successful in creating a number of sites, you have also created a much larger inventory that the prospective buyer must reimburse you for in order to buy the business. Unless you have strong sales numbers, this may cool the prospect's ardor or greatly limit the amount they are willing to pay for "good will", i.e. the difference between the asking price and the value of the inventory. Further if the bulk of the stock is near or at its expiration date, you may be forced to heavily discount the value of the inventory since it is technically unable to be sold once it reaches its expiration date.

XI.
WEBSITE

One of the issues you will confront early on is whether to participate in the company's website. Generally this will be a "replicated website", i.e., a duplicate of the company's website except with your company's name in addition to or instead of that of the Company.

The arguments in favor of participation, assuming you want a website, are:

> Cost—The work is already done and you don't have to pay the development cost.
>
> Time—you are up and running immediately and do not have to lose time working with a website developer. In addition, the difficult task of entering all of the product codes descriptions and pricing has already been done.
>
> Flexibility—You can use the website to offer special discounting to key retailers, enable them to order directly from the site and optimize business-to-business selling.
>
> Updating—this   becomes   the   company's

responsibility and you don't have to worry about re-doing everything when the inevitable price changes occur.

The arguments against participation are:

Term—you may be required to agree to a multi-year contract.

Control—you have no say in the website's design or operational capabilities

Cost—The basic functionality and hosting can probably be obtained a lower cost

Restrictions—Having your own site enables you to sell other products using the same functionality

As exciting as the prospect of a functioning website is, it is important to recognize, however, that a website is only as useful as your marketing makes it so. The benefits of the website include the ability to order items not normally stocked, business to business customers can order direct, commission income with no product handling by you and the ability to set up special discount pricing arrangements for individual customers. These benefits have to be weighed against the need to recover the monthly charge of $40 to $60, the cost of advertising the site in some way, loss of sales due to purchaser's ability to more easily comparative shop for items on the internet.

XII.
PRODUCT ISSUES

---

a.   Expiration Dates

One of the little discussed but most important factors in the ink cartridge business is the role of expiration dates. The ink cartridges have an expiration date printed somewhere on the packaging. Generally it is 18 to 24 months from date of manufacture/remanufacture. While this seems to be an adequate working "window" of time for sale, you cannot assume that you will have the benefit of the entire time.

The cartridges move from manufacture to warehousing and then are put into stock. There may be another intermediary before it is shipped to you the distributor. It may not be a cartridge that is in high demand and may have been on the shelf for a year before you order it.

You should ask the company you plan to do business with to provide you with a statement of their policy regarding expired stock. Some companies will only guarantee that they will not ship anything to distributors that has less than six months left before expiration. As a distributor, there is an argument that if you were provided with "fresh" stock, you should bear the risk of expiration dates since you, and not your supplier, control the opportunities for sale.

While some expiration periods may be adequate in the context of a fast-moving cartridge, it is important to remember that when initially investing in a distributorship you will be sent one or more complete set ups with many different types of cartridges. Some of these will be in high demand, most will not.

Other factors may start to erode your marketing expiration date "window". For example you will probably get your shipment of stock before you have identified sites for your displays. If the site finding process becomes protracted, e.g. 3 months, you have now reduced the six month sales "window" to 90 days.

Expiration dates do not prevent the cartridge from being fully effective beyond that date but knowledgeable consumers will probably not purchase expired stock or you may have to consider substantial discounts. In addition, leaving expired stock on displays has the potential to affect the retailer's reputation and to undermine the credibility of all of the products on the display despite the fact that you might have a strong replacement guarantee. If the presence of expired stock has cooled the consumer's interest in buying, your replacement guarantee becomes irrelevant. And all the while the clock is ticking. Other cartridges on the display are nearing their expiration dates and the already expired stock is getting even older.

The fact is, however, that this is just part of the business and a risk that you have to accept. Two steps you can take to manage the expiration date risk are a) confirming your supplier's replacement policy and b) modifying your stocking policy. Prior to contracting to purchase of distributorship, review the contract to determine the supplier's policy regarding expired stock. What is the time period for return of expired stock? Is it one for one or is there a restocking fee? Is it a refund or credit only? Who pays the shipping costs? The other technique to adopt is to identify the types of cartridges that do not move and carry fewer of them in stock. In other words if you normally carry 3 of each type of cartridge on your displays, consider having only one of the slow-moving cartridges on the display.

Finally it is important to recognize that management of expiration dates will demand additional time and funds. The expired stock has to be retrieved, stripped of price stickers, packed and shipped. In addition, if all of the stock of a particular type of cartridge has expired, you are faced with the choice of not having any of that type available while you are awaiting replacements or you will have to purchase new stock while you are awaiting credit or replacement of the expired stock. The latter approach means that you have additional funds invested, albeit temporarily, and this can affect your cash flow.

b.   Product Knowledge

Unlike bread, soda, snack or other distributorships, the ink cartridge business is complex and has an extensive product line. As a distributor, you bear the burden of becoming a knowledgeable resource for your retailers. That does not mean simply knowing the brands you sell. Rather it involves everything from the principles behind inkjet technology, to cartridge construction, fill volumes, and a host of other issues. In some cases you will need the knowledge to attempt to education the retailer's sales staff. In the case of other retailers, such as computer service establishments, the retailer will assume that you have equal or better knowledge of the cartridges than he has.

c.   Defective Cartridges

Defective cartridges are the bane of the generic cartridge industry. Printer manufacturers do not release data on defective Original Equipment Manufacturer (OEM) cartridges. Even if they did it would probably have no effect on sales. Part of the reason is human nature. If I have a "Brand X" printer and I purchase a new inkjet cartridge manufactured by that same Brand X manufacturer from a "name" store my assumption is that everything will work correctly. When it doesn't I am probably more inclined to have an initial reaction that I might have installed it improperly.

If I purchase a generic brand compatible or remanufactured cartridge and it doesn't operate properly, my immediate assumption is that the cartridge is faulty, especially if it is remanufactured.

Most suppliers have an absolute replacement guarantee for defective cartridges. This is necessary to maintain sales. However, defective cartridges undermine consumers' confidence in the brand as well as in the retailer. The return and replacement is inconvenient because it requires two trips and it may have frustrated the consumer's plans to complete a project at a specified time. Further, it is the retailer, and not you the distributor, who must face the consumer.

Be very careful when relying on statements of low defect rates. For example "Based on cartridges shipped our defect rate is virtually non-existent." This statement is entirely defensible. It is, however, entirely misleading because it compares apples (all types shipped) to oranges (frequency of defects) The only measure that counts for you as a potential distributor is: What are the types (SKU) of cartridges with the highest defect rates and are these types of cartridges that are in high demand? As part of your "due diligence" before you buy a distributorship you should ask for the following:

  a)   a listing of the types of cartridges (not just the printer brands) that will be provided with each display and the quantity of each.

  b)   the suggested retail price and distributor cost of each

  c)   the highest and lowest volume selling cartridges in the display

  d)   the return (defect) rate for each type SKU (not

just brand) of cartridge. For example, Lexmark remanufactured cartridges might have a defect rate of 1% but the most popular type (i.e. SKU) of remanufactured Lexmark cartridge may have a defect rate of 15%. The franchisor will frequent cite a low defect rate which is, in fact, true. However, they may be quoting a rate computed on total returns divided by total sales. This can be misleading.

Consider the sale of 25 cartridges each for four different brands of printers. Assume that Brand Y has 5 defective cartridges returned. Based on total sales (100) the defect rate is 5%. (5/100) But the defect rate for Brand Y cartridges is 20%! (5/25) And if the rate were computed on a particular type (SKU) of Brand Y cartridge, the defect rate might be considerably higher. If Brand Y printers are increasing in popularity the defect rate may sour your retailer on the entire line.

What can you do about it? In reality, nothing. You don't control the manufacturing process and you may be prohibited by the terms of your contract from purchasing supplies from an alternate vendor. Even if you could, the defect rate might even be higher. To some extent the problem is in the industry which manufactures both new compatible cartridges but also remanufactures other types of cartridges due to the fact that they are unable to obtain a license to copy the patented technology built into a particular brand of cartridge. All they can do is refurbish the empty cartridge. These latter are used cartridges. True, they are remanufactured and the quality control may be extensive. Nevertheless, most quality control is based on sampling rather than testing of every item. Even if the problem is with the consumer's printer or the way it was installed by the consumer, the consumer's presumption is often that there must be something wrong with the cartridge since is not an original from the printer manufacturer.

XIII.
BUSINESS PLAN

One of the reasons franchise sales are so successful and attractive is the fact that the organization marketing the franchise or distributorship has seemingly already done the hard work of "crunching the numbers" in order to determine whether the activity will be successful. Just as we all do when we are selling a car or house or other item, the seller tends to show the most optimistic outcomes.

Whether it is sales, turns of stock, or other measure, the numbers always demonstrate your ability to recover your investment quickly.

But you should bring the same level of skepticism to the claims of "business opportunity" marketers and franchisors as you would to the claims of used car salespersons or real estate agents. Their success depends on seeing the optimistic side of everything. Your job is to view the opportunity with the utmost objectivity and reality. And that is not easy to do because of the difficulty in obtaining information from alternate sources.

Nevertheless there are steps you can take to develop a more objective view of the opportunity. The most critical and basic is the preparation of a business plan. In order to get the proper perspective, (even though you will probably be using your own funds) assume for the purposes of this analysis that you will be asking a lender for a loan to enable you to finance the opportunity. The lender would require a business plan. Take the figures provided by the seller and use them to project income and expenses for one year. Then do the same projection several times using a fraction of the seller's claims, e.g. assume sales will be only 50%, 25%, or even 10% of what the seller claims. Remember, you are planning for is the first year of operations for a start-up of a new business, not one that has been established for several years. Consider whether you can afford to wait one to several years to achieve the numbers offered by the seller and whether your return on your investment is adequate at the reduced sales levels you posited.

If you've never done a business plan, don't worry. There are samples and other resources on the Internet that will make it possible to do a rudimentary plan. Remember, you are probably not, in fact approaching a lender for a loan so you don't need a perfect business plan. All you need is a rough estimate of income and expenses to serve as a counterweight to the seller's projections.

Before we leave the issue of projections, it would be well for you to ask for some of the details or studies supporting the seller's sales projections. How big was the database? When was the study prepared? Do current sales reflect those projections? Was the data local to a particular region or national in scope? Was it based on new distributors and retailers or on sites that had been in place for several years?

XIV.
ADMINISTRATION

While marketing issues will probably demand most of your attention and creative energies, be prepared to spend significant amounts of time on administrative tasks. As you think about how much time you have available to allocate to your new business be sure to leave some time for the various mundane tasks and demands of your budding company. The following are just a few areas to think through and estimate how you will accomplish them and how much time it will take.

- Entry of orders for product from supplier into your accounting system
- Entry of sales for each retailer
- Periodic sales tax returns that may apply in your jurisdiction
- Annual tax return information (including possible 1099 forms to be issued to qualifying retailers, if you are lucky enough to achieve the threshold volume of sales that triggers 1099 reporting!)
- Entry of stock purchases and sales of stock in the inventory control system

- Market research for new locations
- Replacement of stock nearing expiration
- Completing suppliers requirements for authorization to return defective products
- Packaging and shipping products
- Development of marketing materials
- Telephone time to answer retailer questions about product and to order unusual items
- Telephone time to deal with suppliers service staff to obtain information/answers to customer questions
- Daily posting of receipts and expenses
- Periodic modification of your systems to reflect price changes
- Possible time to re-price stock already on displays

XV.
ACCOUNTING

As the owner of a business venture you are required to keep financial records in good order and appropriate for the business you are in. That means you will need to purchase an accounting program . You will need to enter each of the types (SKU's) of cartridges along with the wholesale cost and retail price. Think through how you will manage the retailer restocking and invoicing system. Anticipate that there will be price changes and whether or not you will need to re-label all consignment stock with new pricing.

Not all cartridges are the same price. The suggested selling price of compatible cartridge may be one-half the cost of the OEM version and may cost the distributor very little, e.g. one-third of the suggested selling price. This results in an arrangement where the retailer and the distributor each receive one third of the suggested selling price. For example, a cartridge with a suggested selling price of $6 may cost the distributor $2, thus allowing for a $2 "commission" or retainage by the retailer and $2 gross profit for you the distributor.

There may however be other cartridges which because of

scarcity or other reason cost the distributor much more. But in order to keep the retail sale price significantly below the cost of the OEM version, the margins are much lower. For example some cartridges with a recommended sale price of $24 may cost the distributor $16. this leaves only $8 to share between the distributor and the retailer. If your arrangement is that the retailer retains 30% of the sale price of each cartridge sold, the retailer will retain $7.20 leaving just $.80 for you the distributor. If all types of cartridges sell equally, then perhaps you can afford to treat this particular model as a loss leader. However, the more likely scenario is that there may be few sales of the high profit models and significant volumes of the low profit models.

That means that you will have to continually monitor sales volumes and adjust the percentage of retainage, by individual cartridge types, in some cases.

Inventory

Inventory control starts with your decision to invest in a distributorship. The agreement will provide that the company will provide you with a number of displays and initial stock. IT IS ESSENTIAL THAT YOU OBTAIN SOME ASSURANCE AS TO THE EXPIRATION DATE OF THAT INITIAL SHIPMENT. If the shipment contains cartridges that have a expiration date of, for example, one year from the time you receive them, you will be under pressure to get them sold prior to the expiration date. Remember, however, that if you have decided to use the company's location services they may take several months to locate sites. Your company may have a rule that they will not accept any expired cartridges. That means that after about 4 months you will have to start taking back the initial stock so that you will be able to return it before the expiration date. The company will probably have a "restocking fee", e.g. 15% . It may also have a rule that all returns have to have the price stickers removed or a fee charged if the company has to do it. Return of slow moving stock also means that you may have to order more stock so that you can re-stock the displays. Do no assume that you will be able to do a simple and timely exchange since the company may have a rule that they will not issue a credit until the return merchandise has been inspected. That may take thirty days or more. Develop a system for listing stock levels for each retailer (they may be different). You will need another system for tracking your own inventory used for restocking retailers' displays and for re-ordering stock. Finally, inventory ties up your cash. You will have to find a balance between how many of each type of cartridge you will put on each display and how frequently you are willing to visit retailers to restock.. The company may urge you to have a particular level, e.g. six per item, or more or less depending on the type. The fewer the cartridges the more frequently you may have to restock but the lower the amount of your cash is tied up in inventory. The company and the retailer may argue that a heavily stocked display looks better. That may be true, <u>but it's your money</u>!

XVI.
ANCILLARY BUSINESS ISSUES

Type of Entity

Check with your accountant and /or attorney to determine whether you should consider doing business as a corporation , a limited liability company or simply as an individual. There are benefits and burdens (taxes, liability, salability, etc) to each form of entity.

Space

Allocate space in a heated area for upright storage of cartridges. The cartridges have to be protected from freezing and from extreme heat. Set up a system of inventory control that will move the stock with the shortest expiration dates to the front.

Mobile Operations

You will need a system for carrying a supply of cartridges so that you can re-stock sold items. Assume that it will be raining on the day you re-stock. Do not rush out and buy a van. You will probably be able to restock ten or more retailers with a car and some shallow (e.g. under bed sized) plastic storage bins that are labeled and can hold several of each type of cartridge segregated in each bin.

XVII.
CONCLUSIONS

Part of the attraction of inkjet cartridge distributorships is the consumers' overwhelming dissatisfaction with the current "cheap equipment/expensive supplies" system is the fact that it is inherently counter intuitive. Presumably, ink is the lowest cost item in the system that consists of a printer, a container for holding the ink, and the ink. At this point, however, depending on the type of printer, an OEM ink cartridge could cost as much as one third of the cost of the printer. The system is illogical on its face and therefore the consumer welcomes an opportunity to replenish ink at a lower cost.

This might lead one to suspect that a system that offers cartridges at one-third to one-half of the cost of comparable OEM cartridges would have the product flying off the displays. And certainly the folks trying to sell you the distributorship would like you to think so.

Sadly, that is just not the case.

• Despite federal legislation that prohibits manufacturers from voiding warranties of non-OEM parts are used, the consumer remains concerned that using non-OEM parts will create problems.

- The printer manufacturing industry is constantly introducing new printers. While these may use the same type of cartridge that is already in use, in many cases the new printer uses a new type of cartridge. That means that you will not have it and that the industry will not be able to provide them to you until there is a sufficient supply of used cartridges of the new type obtained by the remanufacturers. (In the case of printers that can use compatible cartridges, there may also be a delay while the manufacturers of compatible cartridges "reverse engineer" the new cartridge type or otherwise manage to develop a compatible cartridge for use in the new machine.

- These new types of generic brands may not be considerably less expensive due to the high development costs in the case of compatibles or due to the cost of obtaining used cartridges to remanufacture.

- As with any product, there is the potential for defects. The cartridge industry does not test every cartridge it produces but relies instead on sampling. Where most aspects of the production process are controlled, as in the case of the production of *compatible* cartridges, sampling is appropriate and effective. Where however, the process involves the use of a used cartridge, the chances for defects increase considerably. While the industry offers a replacement warranty, the inconveniences associated with obtaining a replacements tend to undermine the consumer's confidence in the entire aftermarket industry as well as the retailer's confidence that carrying the product will support the retailer's quality image.

- As previously mentioned, it is doubtful that the independent distributor will be able to find outlets that will be able to match the convenience and advertising awareness associated with the major office supply stores.

Market, Market, Market

What's the point of all of this? Simply that the hallmark of most distributorships is not primarily order-fulfillment. Rather it is constant marketing and sales. You cannot rely on the retailers to push the product since the volume and unit price of the cartridges do not warrant the expenditure of the retailer's resources. Instead, you will be responsible for driving sales. *A decision to purchase an ink and toner distributorship is a decision to go into sales.* You have to sell the concept to retailers. You have to alert the consumer to the availability of an alternative to OEM cartridges. You have to educate the retailer and consumer about the benefits of alternative ink products. In short, plan on spending at least one-half of your time on marketing and sales. Is it possible to become rich as a single distributor? Perhaps. But the real question is: What is the potential return in another form of business where you would commit the same amount of time, energy and capital?

When you decide to purchase a distributorship, remember that you are investing in someone else's dream. And the essence of that dream is likely the sale of distributorships, not the sale of cartridges.